Beginning 2022–06–27

2024 Edition

I0422364

,,,

THE HISTORY OF THE PERPETUAL MOTION BOMB

by Nathan Coppedge

,,,

BIOGRAPHY AND INTRODUCTION

Nathan Larkin Coppedge (b. 1982) has done extensive research on perpetual motion. Beginning in 2009 Coppedge dreamed of applying the 'unlikely' devices to defense applications. For example, he thought of a static electric staff that would recharge electricity automatically, and a way where if the machines worked, they could be used to recharge a laser battery, or perhaps with later developments, even be used to create some type of over-unity explosion. These explosions are currently thought to be make-believe.

However Coppedge is interested academically from the standpoint of promoting perpetual motion machines. Coppedge has never been employed in government except as a public library assistant. This research comes almost exclusively from his own experiments since 2000. It is mainly focused around over-unity type effects, nothing similar to weapons used by the military. Scientists currently believe that these devices are not physically possible as they involve a phenomena which are largely considered impossible to science as of 2024.

,,,

THE HISTORY OF THE PERPETUAL MOTION BOMB

"It would say too many booms let's put it that way." —Unknown expert

Note: This is not a real government project as far as I know. The technology is thought to require perpetual motion chemistry, which is currently regarded as 'more impossible than impossible' by most scientists (chemistry tends to be metabolic, not mechanical).

(7) You could just try fake Perpetual Motion Bomb: there may be alternatives but they are not as good as the procedure, (8) Carriable Perpetual Motion Bomb: the perpetual motion bomb is normally bulky, so the ability to carry the bomb is a factor, (9) Improved Perpetual Motion Bomb, it may be important to research perpetual chemistry, which is the factor that makes the bomb continue burning perpetually (10)(a) Adjustable Perpetual Motion Bomb: It may be important to have a step where the bomb is compacted or activated prior to using the bomb. (b) More effective Perpetual Motion Bomb: we have created hell. (c) Adjustable and more effective: The eternal fire or endless storm. (d) Monster Perpetual Motion Bomb: We have created the Big Fire. (e) Carriable Perpetual Motion Bomb: Throw a fireball. (f) Industrial Perpetual Motion Bomb: Fire seeds, the bombs could be used as an exciting new fuel source

Perpetual Motion and Military Defense Links.

(almost a joke weapon) <u>Allaying Fears of Nuclear Armageddon Resulting from Perpetual Motion Machines</u>.

Nathan wrote around 2010: The gist of this is, that perpetual motion may be considered better than magic if it works: principle of perpetual motion defense.

Disclosure: It is thought perpetual motion bombs are impractical, at least according to conventional science. Perpetual motion is more like swords to plowshares than warfare anyway. Most perpetual motion weapons are likely to be defensive rather than offensive, though energy from perpetual motion could potentially be used for applications like defensive lasers or industrial development.

2001: Dinko M may have theorized that an Alcubierre drive might destroy the known universe. Thus, the concept of something like cold fusion has been linked to something involving destruction of planets.

The main theme here however is over-unity prior to any type of explosion or melting.

2009: Nathan Coppedge theorizes that perpetual motion chemistry may exist.

2010: "The gist of this is, that perpetual motion may be considered better than magic if it works: principle of perpetual motion defense." —Nathan Coppedge

August 19, 2020: Writing on <u>Perpetual Motion Machines of the 2nd Kind</u> (...)

CHEMICAL PERPETUAL MOTION EXAMPLES (PURELY THEORETICAL):

Possible examples based on a general elemental equation:

Lithium - helium.

Hydrogen annihilation using antihydrogen.

---Exploring Chemistry

As far as human immortality:

Lithium + anticarbon + byrillium

Hydrogen + anticarbon + boron

--Telaletheian Science

August 21, 2020: Writing on <u>Chemical Perpetual Motion of the 1st Kind</u> (...)

PURELY THEORETICAL NO MONEY IS BEING TRANSACTED HERE

Elements - 1 = Perpetual Motion (ordinary difference + 1), this could also be seen as knowledge of elements.

Helium - hydrogen = super-perpetual, this could be seen as elementary knowledge of helium.

Elements after lithium - hydrogen = exemplary advanced perpetual motion, this could be seen as elementary knowledge of elements after lithium.

Antihelium + lithium = exemplary perpetual motion, perhaps related to advanced flying.

---Exploring Chemistry

FOR HUMANS:

Elements - 1 = Perpetual Motion (ordinary difference + 1), this could also be seen as knowledge of elements.

Carbon - 1 = Perpetual Motion (ordinary difference + 1), this could also be seen as knowledge of carbon, hydrogen passing upwards through carbon.

Helium + anticarbon + boron = super-perpetual, perhaps related to flying.

Elements after lithium + anticarbon + boron = exemplary advanced perpetual motion.

Antihelium + carbon + antilithium = exemplary perpetual motion perhaps related to immortal languages.

---Telaletheian Science

POSSIBLY USEFUL INTERPRETATION: Minus = Lighter elements leaving gravity. Plus = Lighter elements attracted by gravity.

2021-07-01: Basic Energy Equation:

FOR ORDINARY OBJECTS

$$[(MIN\ EFF + 1) - (MAX\ EFF + 1)] / [0.5\ (MIN\ EFF + MAX\ EFF)]$$

--> Equivalent to: Min Eff - Max Eff / Avg Eff

FOR PERPETUAL MOTION:

$$[(MIN\ EFF + 1) - ((MAX\ EFF / 2) + 1)] / [0.5\ (MIN\ EFF + MAX\ EFF)]$$

It is shown if the formula for ordinary objects is correct, the additional modification of Max Eff / 2 simply reflects the use of a counterweight versus a gradient, which is not a large modification.

[In a perpetual motion bomb a chemical equivalent of the mechanical process would have to be effected, which is not known to have occurred in 2024-02-07]

If you subtract 16% for friction, the maximum over-unity decreases to about 134% unless you introduce an additional principle like lighter-than-air or an additional device adding energy.

Some theoretical devices predict higher than 150% but there is not much evidence for them yet. In fact, cycles in general have been very hard to create so far, but research is ongoing. —General Over-Unity Equation

2022-06-27:

"There is a certain cantaneelish appeal to perpetual motion: I may have already been worth more net energy than atom bombs... All without paying for the Manhattan Project..." —Nathan Coppedge (Cantaneel Emily Dickinson 0 Relevant Results)

This prompted the posting: Scariest News Prior to June 21 - 27 2022

2022-06-27:

EXTENSION OF EINSTEIN FOR PERPETUAL MOTION BOMBS AND SIMILAR: Substitute C^2 for Difference in NC Tree Theory formulas and you may get new nuclear formulas. A Simpletonian View of Einstein on Differences 2022-06-27

2023-05-21:

Diagram drawn up. There were later revisions to some of the other devices. The operation of chemical perpetual motion is still not fully known to my knowledge in December 2023.

OVER-UNITY WEAPONS

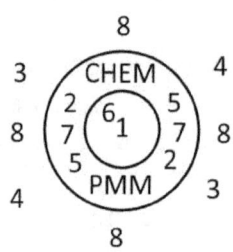

PERPETUAL MOTION BOMB "ETERNAL FIRE"

CRITICAL IS THE PRIOR
INVENTION OF CHEM PMM.
OTHERWISE, THE INVENTION
ACTS THROUGH PURE
CHEMISTRY EXCEPT
WITH AN ENDURING CORE,
AND REPEATED EFFECT
WHICH DEPENDS ON
EXTERNAL HEAT ACTING
ON THE OUTER CORE

CHEM PMM MAY ALSO DEPEND
ON THIS SORT OF REACTION
BEFORE EXISTING ITSELF
OR IT MAY NOT.

TO REIGNITE THE CENTER,
THEN CAUSING A MORE
EXTREME REACTION WITH
CHEM PMM WHICH IS THOUGHT
TO BE SUSTAINABLE ONCE
CHEM PMM EXISTS.

SELF-RECHARGING LASERS

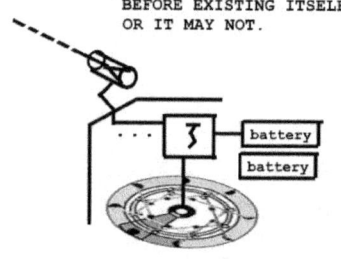

ONE OF THE FIRST 'IMMORTAL
DEFENSES' EVER CONCEIVED,
THE RELATIVE PERMANENCE OF
A WELL-BUILT LASER WHEN
COMBINED WITH AN PERMANENT
ENERGY SOURCE YIELDS LASERS
WHICH REQUIRE LITTLE TO NO
UPKEEP AND WHICH PERIODICALLY
RECHARGE THEIR ENERGY, OR CAN
STORE LARGE CAPACITIES THROUGH
DOWNTIME.

ESCHER RIFLING

AT LEFT IS AN APPROXIMATION
OF THE 1.1025 PERCENT DEGREES
H X V NECESSARY TO CREATE
AN OVER-UNITY GUN USING
CONVENTIONAL PROPELLANT.

IMPROVED DESIGN **LEVER**
GAS FROM
PROPELLANT

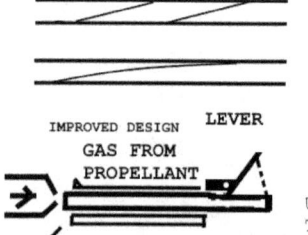

DIRECTION   THIS WIRE
OF RECOIL MAY BE IMPORTANT
BARREL SLIDES SL.

KINETIC RECOILLESS RIFLE

USE OF AN ANGULAR CONNECTION TO
THE BACK OF BARREL COMBINED WITH
FORWARD LEVERAGE ON PROPELLANT MAY
CREATE DOWNWARDS RECOIL OR NO RECOIL,
AT ALL THROUGH USE OF A PRINCIPLE
SIMILAR TO THE EXP GRAVITY LEVER.

ENERGY ARMIES

AN ARMY WITH AN UNLIMITED ENERGY BUDGET USING
PERPETUAL MOTION COULD POTENTIALLY BE OVERWHELMING.

"FREE-WEAPONS"

INCREASING A MILITARY'S ENERGY BUDGET MAY CREATE RESULTS
SUCH AS INCREASED CONVEYANCE, INCREASED WEAPONS
PERFORMANCE, QUICKER ENGAGEMENT, AND IMPROVED
ECONOMICS OF WARFARE.

2023-06-20: Nathan Coppedge writes on Reactive Mechanics, part of the theory of Reactive Mechanisms:

BASIC MECHANICAL REACTIONS

'MATH REACTION'

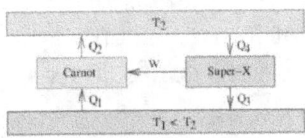

In the above diagram we can see the hypothetical 'super-x machine' as being equivalent to some special electronic process perhaps involving some type of overclocking or mathematical process using exponential efficiency.

'CAUSAL REACTION'

A series of elements, ideally each of them self-resetting, is set up so as to cause particular effects in each unit through an initial impulse. Mathematics or splash properties could be used in each unit so as to enhance the effect of the reaction.

'RISE AND SPLASH REACTION'

Using mechanical properties, energy might increase periodically, leading to a key major reaction using additional enhancement or chemistry. In this type, ideally the major reaction is charged periodically by the rising energy using mechanical properties or stored energy.

'SPLASH REACTION'

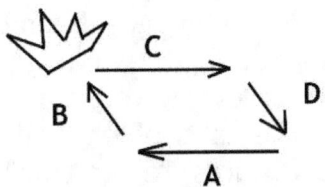

A basic chemical perpetual motion machine might not use a mathematical process, instead it might use something like flint-and-steel, with an object striking against a particular surface periodically. The type of reaction is not particularly important as long as it does not require much energy. What is important for the concept is that the cycle occurs periodically, involves some type of visible or audible reaction, and is repeatable. Probably the use of mechanical properties to create the reaction is ideal, in which case it is not inherently chemical at all. (For example, 'splash').

'ACCUMULATED REACTION'

Energy could be siphoned out of the reaction into an underlying field, which stores the energy, and feeds some quantity back into the system.

'MULTIPLIED REACTION'

Pictured at right, if a device is permitted to run in a channel, with other machines running parallel with the same direction of motion, this may be used to enhanced some of the other effects, and to create an enhanced reaction in the center of the path of motion.

Nathan Coppedge
2023-06-20
Publicly released.

MATH REACTION DIAGRAM BASED ON:
https://phys.libretexts.org/Bookshelves/University_Physics/Radically_Modern_
Introductory_Physics_Text_II_(Raymond)/24%3A_The_Ideal_Gas_and_Heat_
Engines/24.03%3A_Perpetual_Motion_Machines

...

August 28, 2023:

An article has been released on a possible anti-Escher Machine gravity anomaly which occurred around late 2023-08-28: "A gravity problem on Aug 28, 2023 may have been caused by some authority overreacting to working perpetual motion or more like something very similar… Note very confusingly this critical problem had to do with the [downwards] return track… cosmic disaster [?]"

Ongoing problems with gravito-geometry:
Someone should research whether spheres still roll downwards. There may be a 'weird gravity polarity' that doesn't make sense that is very supposedly preventing gravity from working anywhere in the universe. Earlier, may have overreacted with the "Expect orbital problems soon." warning. In any case, watch out for ignoring the properties of combined machines and / or
standard gravity where conditions apply

—Article on Gravity Anomaly of Aug 28, 2023

2023-12-06:

The perpetual motion nuclear bomb and bouncing nuclear bomb were general concepts by this point, as variations on perpetual motion weapons (though they are probably not taken seriously by government, as the technology does not exist, and may not ever). There may be a similarity to the Chinese 'artificial sun' concept.

...

2023-12-12:

Possible instructions on designing perpetual motion weapons: OU Load Out (google docs). Description of development using program.

...

2024-01-01:

One theoretical weapon which may have higher destructive potential than Earth's scale is the highly- theoretical concept of 'weapons of distance'. These devices would use over-unity properties to create an exponentially-increasing explosion which could react between multiple stars, destroying what might be a large part of the visible universe in one explosion. However, scientists consider over-unity weapons impossible currently, as publicly at least scientists have not believed Nathan Coppedge's claims to over-unity.

However, some other much weaker over-unity weapons may be easy for militaries to build, and have some supporting evidence, including self-recharging lasers, Escher rifling, over-unity recoil, forward-fire ability, the perpetual motion magical staff, and perpetual motion-powered traps.

2024-01-03:

Nathan Coppedge thinks a perpetual motion bomb could be built by universities, but more likely national governments. The processes necessary to build such a bomb are not yet realized. Perpetual motion chemistry remains the key fundamental requirement other than bomb technology. Technically this may be the invention of modern university warfare. Another discovery at this point is the realization that perpetual motion bombs may require a third component involving a cyclical process, perhaps involving a chemical-thermal computer or external wireless control. It is also possible that perpetual motion bombs could be made in conventional and non-conventional varieties, thus serving as a gap between ordinary bombs and fissile and fusile bombs. Though it is also noted perpetual motion bombs are not likely to particularly contribute to nuclear weapons technology as their unique components are an addition to any existing weapons system, a method that is familiar to those mentally conformized to applied perpetual motion technologies.

2024-01-03:

Perpetual motion bombs, like other perpetual motion systems, may have unique properties which are unique to perpetual motion-related research. Some of these properties are concomitance (same device repeats), perduring destruction (repeated damage to one area), in the worst cases heat envelopment (heavily-overlapping destructive effects over noticeable periods of time in one area), inexorable destruction (permanent uncontrollability mainly affecting one area unless it is a bouncing or traveling bomb), and eternal destruction (perpetual re-ignition). A key discovery at this point is the traveling bomb similar to a regenerating exploding unmanned tank which mixes offense and defense.

...

2024-01-14:

It was discovered that the mildly related antigravity bomb may involve Plutonium. This may be the first mention of the antigravity bomb.

...

2024-03-01: Minor development: THEORETICAL OVER-UNITY JETS, Theorized since 2009: Ideas for future airliners: 1.1025 degrees twist flat spiral similar to rifling may allow faster jets (not tested at all yet that I know of). Over-unity jets. The multiplier is a similar 110.25% on velocity as far as is known, with some evidence. Tilt motor (not tilt rotor) once proven, with heavy rolling cone may permit faster rockets if attached contingently and horizontally. Two could be put above rocket and two below to create lower heating and up to maybe a X6 multiplier on fuel efficiency minus some extra weight of about 4X (scalable to the fuel efficiency). However, unless designed correctly it could be extremely dangerous and somewhat rocky to increase the speed of rockets, so I usually assume it will be applied first in the military. The real fuel efficiency increase is only up to 34% of the mass of the rolling cones expressed as propulsion = 134% multiplier on fuel efficiency, with 100% expressing the existing range. Note that I am not an expert on rocketry or aerodynamics, so treat all this as very very tentative and untested. —Theoretical Over-Unity Jets

...

2024-03-02: All Possible 'Efficiency' Equations Modified for Over-Unity:

Theory of Everything: [(MIN EFF + DIFF) - ((MAX EFF / 2) + DIFF)] / [0.5 (MIN EFF + MAX EFF)] + Diff X 100%

Negative Theory of Everything: [((Max Results - Min Results) / ((Max Eff / 2) - Diff)) - Diff] X 100 (%)

Anti-Theory: [(Difference - Eff) - ((Difference / 2) - Eff)] / [0.5 (Min Diff + Max Diff)] - Eff X 100%

Disintegral: ((Max Results - Min Results) / ((Max Eff / 2) - Diff)) - Diff X 100 (%)

Efficiency Formula: [(Results - Diff) - ((Results / 2) - Diff)] / [0.5 (Min Results + Max Results)] - Diff X 100%

Anti-Efficiency Formula: [(-Results + Diff) - ((-Results / 2) + Diff)] / [0.5 (- Min Results - Max Results)] + Diff X 100%

Difference: [(Results - Eff) - ((Results / 2) - Eff)] / [0.5 (Min Results + Max Results)] - Efficiency X 100%

Anti-Difference Equation: [(-Results + Eff) - ((-Results / 2) + Eff)] / [0.5 (-Min Results + -Max Results)] + Efficiency X 100%

Forces (Technology): [(Antiforces + Forces) - ((Antiforces / 2) + Forces)] / [0.5 (Min Antiforces + Max Antiforces)] + Forces X 100%

Negative Forces (Technology): [(-Antiforces + Neg Forces) - ((-Antiforces / 2) + Neg Forces)] / [0.5 -(Min Antiforces + Max Antiforces)] + Forces X 100%

Antiforce Equation: [(Antiforces - Forces) - ((Antiforces / 2) - Forces)] / [0.5 (Min Antiforces + Max Antiforces)] - Forces X 100%

Negative Antiforces: [(Neg Forces + Forces) - ((Neg Forces / 2) + Forces)] / [0.5 (Min Neg Forces + Max Neg Forces)] + Forces X 100%

Dimensions: [(Dimensions - Antiforces) - ((Dimensions / 2) - Antiforces)] / [0.5 (Min Dimensions + Max Dimensions)] - Antiforces X 100%

Negative Dimensions: [(-Dimensions + Antiforces) - ((-Dimensions / 2) + Antiforces)] / [0.5 (-Min Dimensions + -Max Dimensions)] + Antiforces X 100%

Anti-Dimensions: [(Dimensions - Forces) - ((Dimensions / 2) - Forces)] / [0.5 (Min Dimensions + Max Dimensions)] - Forces X 100%

Negative Anti-Dimensions: [(-Dimensions + Forces) - ((-Dimensions / 2) + Forces)] / [0.5 (Min Dimensions + Max Dimensions)] + Forces X 100%

22

—<u>Core Pinnacle Technologies</u>.

...

2024-03-08—<u>Eternal Fuel</u> : An adaptation of chemical perpetual motion, it is added that chemical perpetual motion could be used for rocket fuel, creating eternal jets. This is assuming the perpetual motion innovation known as eternal flames, a development of perpetual motion bombs, could be used for this purpose or improved for higher efficiency. Another adaptation is that the immortal rocket fuel might be used in outer space, permitting connection to 'lost planets and lost solar systems'. Note also that mechanical efficiency similar to Escher Rifling and OU Recoil might be combined with Eternal Fuel to create more effective flight inside and outside the Earth atmosphere. The invention was prophecied in 2006: *While they were flound'ring / I was pond'ring... / No more wand'ring... Thro' the dark tunnels of grim determination. / For no, it is time we grow in a thousand-folded folds for which we need an infinite fuel.* (Nathan Coppedge)

2024-03-08: Eternal Fire Plasma Weapon:

It is thought a new development of Eternal Fuel concepts is to use eternal rocket fuel methods to power some type of energy weapon. One variation of this is to use 'contained and pressurized high energy fuel' combined with the over-unity levels of energy native to perpetual chemistry to permit 'periodic outbursts' with higher power than conventional ammunition... The generation of actual plasma might be a separate factor, as currently it is not known if perpetual chemistry can generate plasma at all.

In fact the mechanisms for perpetual chemistry is currently not known... [The] math for some concepts has sometimes suggested ratings [maybe] as high as 134% ... almost all devices other than N. Coppedge's are thought to be utter failures with ratings rarely higher than 1% of unity.

--<u>Eternal Fire Plasma Weapon</u>

BOOK RECOMMENDATIONS

THE BOOK OF THE FOUR

THE PHENOMENAL HISTORY

THE NECESSARY PERFECTIONS

THE BLACK SWAN MARKET

100 GREAT PERPETUAL MOTION MACHINES

*50 GREAT FLYING AND UNDERWATER PERPETUAL
MOTION MACHINES*

THE HISTORY OF COHERENCE

THE HISTORY OF PERPETUAL MOTION MACHINES

NECESSARY SYSTEMS

SCIENTIFIC THEORIES

THE ALCHEMY

THE SYSTEM OF ALL SYSTEMS

Bio

Nathan Coppedge or Nathan Larkin Coppedge (b.1982) is a philosopher, artist, inventor, poet, and member of the international honor society for philosophers. A prolific author with over 200 books published on Amazon, he is a perpetual motioneer, famous quotable, and internationally-selling Hyper-Cubist. A one-time member of Tesla Society UK online and PESWiki, and founder of many Facebook groups, he lives near Yale University.